Hello.

ENDORSEMENTS FROM:

My Njardu sister-gurl, Violet and Njardu niece Nuchie.

Uncle (my Njardu 'boss-woman's' husband) who still survives her and who is my good mate and tribal brother.

Nuchie said I've thrown a few spanners around all over the place. We can't all wait to get out bush again.

FOREWORD

There's a few crew who I wish really, really would read this book and these are;

All Indigenous health workers and doctors, and

Anyone else involved in the delivery of Indigenous health care programs and their consequent delivery, and

All members of the Australian Traditional Medicine Society (ATMS) and also ANTA and any other National credentialing bodies for naturopaths and other natural therapists, and

All the officers of those Associations, and

Students who may join those Associations, and

People involved in Indigenous Tourism in any way, and

Chris James, and other people delegated to further the cause of Indigenous reconciliation business, and

All 'New-Age' mob and other people interested in spiritual stuff, and

Maybe a politician or two, and a few activists, and

Botanists and other people whose professional life brings them into contact with the Aussie bush, and

Chemists and representatives of pharmaceutical companies, and

Doctors interested in Indigenous issues relating to health, and

Some people involved in the media (journalists) and

I'd also like to think there would be a couple of copies out on every Indigenous remote community, and

finally YOU.

This little book has basically only one purpose which is to explain who I am, and how I got to be in 'this situation' I find myself in, and some stuff and concepts that I'd like anyone interested in learning more about traditional 'earth-based' and bush medicines to know about and understand.

Then once you've gotten your mind around these (ground level) things then we can progress further from there.

You can read it online or if you want to be real flash you can order a printed copy and wait until it arrives from the USA. I've found that often such things might not take very long.

Who knows where this journey is going to take us!
I feel like it's an exciting beginning.

SECTION ONE;

The first section is ABOUT ME and explains to you who I am, where I came from, some places I've been and how I got to be at this point right now.

SECTION TWO;

The second section is more directly about BUSH MEDICINE and the issues that surround it.
But no way do I want you to start reading from there only cos a lot of the stuff in the first section is real important.
And the two go hand in hand together.
This book is written particularly for my Indigenous crew and I make not too many apologies if some of you whitefellas find it a bit tricky to read or understand. Cheers.

Now let's go down to things.

ABOUT ME

I was talking to Uncle the other night and said, "I better write a book in a hell of a hurry".

I told him I was gonna call it "piggie in the middle" which made him think about the days he was working in the missions and about those kids taken away from their folks and he called them 'piggies in the middle too'. Uncle's also a piggy in the middle; his dad was a 'mainstream' white New Zealander and his mum was a 'moldy'. Then he married a full-blood Njardu woman and became piggy in the middle in a wide range of her cultural dealings and affairs.

I'm a piggy in the middle for different reasons which I will get round to explaining to you but just give me a bit more of your time first And don't go throwing rocks or *anything else* at me please until you've read my story. After you've read it and I know that you've heard me then let's then sit down and yarn and you can have a go at trying to tell me what I should or shouldn't do.

It all blew up and out into the public eye the other day when I put some stuff up about Bush Medicines onto Facebook. I always knew it was coming some way or other sooner or later.

I've talked to a few people about it too.

When talking to one of my Indigenous daughters last night she said, " you better write the book in a hell of a hurry and write it out just like you're yarning to me here at the dinner table".

I decided she was right so this is exactly how I'm gonna write my tale to you.

Another of my Indigenous daughters said, "good on-yer mum – give me a box-full".

My non Indigenous daughter said , "hm-mm mum – you know how to start a good fight".

Another member of one of my extended blackfella families was real shocked and shamed and we had a pretty long debate. But she came right later.

One thing she said and these words are really hitting into me were, "Indigenous people will never get any sense out of a Government Department. They will always set up our mob for failure". I think she might be right and I'll go into that more a bit later on too.

She told me that bush medicine was the only thing left that was theirs and that it could not be *stolen* too.
Guess I better tell you now bout how I got to be piggie in the middle.

I think this story *might* have started lotsa years back when I took up a job in a maximum security prison as a teacher in the Indigenous Education Programs.
I say *might* cos really I can think about some other things that have happened in my life that might move this start date back farther quite a lot.

There was this black, black Nigerian fella who was in charge of the "Indigenous Education Programs". He told me,"respect their culture, work within it and don't do anything to hurt or damage it".
He engaged me becoz he thought from my resume that I could run some real interesting courses. So he officially

engaged me to come in and teach "medicine" at the same time as meeting literacy outcomes.

Before I actually arrived into that prison he went around to all the prison units to advertise the fact that he thought he'd done the right thing culturally by engaging me who he thought could get *them* interested in education and classes.

Now let me mention that this was a man's prison and last time I looked I definitely wasn't one.
And I wasn't one back then either.

SO; my first class in that prison was ..well how the hell can I put it....frikking interesting! No other way to put it. The most senior desert lawman in the prison started the ball rolling in my 'interrogation'. I knew completely nothing about lawmen at the time.

I did notice that the men were clustered into small groups and that the oldest did the asking of any questions.

From memory the first question was, "where we go after we die"?

Walking through a large prison with a cazillion gates with armed and mean looking guards was one thing that people could find intimidating but nothing prepared me for this; 'my interrogation'.

Other questions asked were;
You believe in God Miss? How many children you have? How old you are Miss? What stuff you cured? You see spirits? You know any Aboriginal people?
I think I was pretty lucky to have gotten past that first question – all the others seemed much easier.
About ten minutes into my interrogation one of the old fellas put his arm into the air and several of them chimed in, "she's a dreamer – she's right".

And they were impressed that I didn't 'do' doctors and had cured just about anything they could think of. All without the help of the mainstream medicine system.
I think I'd passed the audition. And I even had some Aboriginal friends. I guess they decided to give me a go.

I'd had a pretty interesting cultural background and had spent my childhood living overseas and mixing in other cultures. Not much intimidated me culturally and I felt surely I could do a good job. When I left that prison that day something inside me had changed.

I soon learned how to get along with these fellas (much to the surprise of other teachers within the prison system) and soon had tribal fellas meeting educational outcomes as verified by education moderation meetings.

There were two main older fellas who came from different regions; Western Desert and Kimberley and they were always waiting at the gate before I got to class. They were both 'medicine men'.

And they were eager to share culture too. And they thought I had the nouse to not fuck things up. They considered me a good and trusted friend.

We shared together many of our medicine ways and I soon learned that medicine, law and religion were one and the same inseparable thing.

And that women could not teach men certain things. Not that I ever tried. Though I got special privileges because I had gray hair and I often got to become piggie in the middle negotiator for them like a bridge over troubled waters.

That was the beginning of my prison career and I got hired at many prisons and even did research framed to make the mainstream education system far more accessible to our Indigenous brothers and sisters.
A couple of old indjibarndi women in Bandyup Women's Prison gave me the name 'Jaahda Jinnah' which has stuck now for many years. Looking back on it I think they were from Jigalong. And a few of you's would know what this name means.

I'm getting round now to telling you about my background.

Many Indigenous families have claimed me. I've been out bush a lot. I've been around a bit traveling. I joke sometimes that if I stand in the street of any country town in WA for a little while that someone will claim me.

Which they do – then take me home, treat me like a queen, give me bacon and eggs every morning for breakfast and the best bed in the place.

So I've got around a handful (or two) of families who claim me. And I've had (got) a couple of Indigenous husbands but I'm definitely not going into that here. Once a husband always a husband; a completely different story.

None of them Nyoongars whose boodja I'm currently living on though – but I've got some strong alliances and mob who will talk up for me.

I just don't know where to start. There's a maze of wonderful and interesting coincidences.

I studied Naturopathy (Natural Medicine) in the waybacks and learned all sortsa interesting stuff that has kept me and my family out of the hospitals and doctor's offices.

I ran herbal clinics for many years. I was Goldfields Herbal Clinic for a few years. On Burt Street, Boulder. About one month before she died (I was thirty) my mother told me that she too had done a college Diploma in Herbalism.

This was a couple of months before I finished my course and it spun me out how she left it so late to tell me being that I had been studying for nearly four years then. She also told me she had not been allowed to practice as my dad was in the RAF.

But looking back on it I should have known. She was an unofficial pharmacopeia and medicine woman who knew all kinds of things about what was in our foods and about medicine plants. And she became a very famous psychic clairvoyant.

There was always gazillions of visitors at home. And thousands at her funeral.

She told me stuff about ancestry but it mostly went by me. Buggar. Now I've had to try and reconstruct things and when money and time permit together I study my genealogy and then some of my mothers words come back to me.

She told me she had some 'dark blood'. Spanish which was about as respectful as she could make it sound at the time. She also told me we had 'blue blood' and I have managed to track that coz if you are related to royalty there's already lots of genealogy work done for you by others. I've got a direct line to Robert the Bruce – my 22ggrandfather.

With all kinds of weird and interesting mob
in between. It turns out we've got ancestors from all over the damn place.

I didn't know any of my nanna's or pops and aunties and uncles and hardly any of my cousins. Being in the RAF meant you didn't have much to do with them and my mother was mostly brought up in an orphanage (it seems likely that her mum was murdered in a domestic violence incident) and my father's mob were equally mysterious and I've found a few skeletons in that closet too. Shame and scandal in the family. She told me we had Irish, Scots and mysterious 'dark blood'.

I remember nearly getting expelled from school when I was around ten or eleven for leaving the school at lunchtimes to go 'mix with the gypsies'. This was a very serious affair. My mum was summonsed into the school for an official appointment. I had to sit quietly by. She got real annoyed with the Headmaster and stormed out and I wasn't suspended or expelled.

My mum used to often sing a rhyming ditty that related to her paternal Irish surname about the origins of Irish gypsies.

But something else mysterious happened. Back around the same time I started working in them prisons.

I'd run herbal and natural medicine clinics for many years but really I was never any good at earning a living even though I'm pretty good at medicine business.
I never made a profit or even enough to survive. I always

made enough money as a well known clairvoyant and doing part-time teaching jobs and this managed to keep my medicine business going.

One afternoon a lady came for a reading and was sitting in my practice room. As I was doing the reading my mother's spirit appeared. At the time my mum had been dead for around fifteen years.

I asked this woman, "Why is my mother standing next to you"? "Ah", she said, "is your mother Chris Johnson"?

She went on to say she had been looking for me since my mum's death and that she had messages for me. Sure I was intrigued.

She told me that a few months before my mum's death she had been visited frequently by desert women from around Alice Springs who had asked my mum to help in some way or other with medicine business and (apparently) my mum had told them that she was gonna die soon and that her oldest daughter would help.

Very strange. Interesting I thought.

Pity she didn't tell me this but if stuff's important it comes back round. I was living on the other side of Australia at the time.
This woman also gave me a couple of other messages that related to Indigenous matters. Not relevant here. I think.

I've written articles on a popular internet site that pays me a couple of peanuts every time I publish something there.

One's about getting struck twice by lightning in one afternoon and another one is about some of the main differences in learning styles between mainstream and Indigenous people. Two-ways business.

I'm a bit prone to getting struck by lightning. Well should I say that I personally don't get struck – the house or the car I'm in gets struck. It's a pretty unique kind of experience –very very loud and very very bright. And a bit smelly too.

It tends to happen just before my life is going to change big time. One afternoon I got struck twice. I had driven through the desert from Mount Magnet through Sandstone to Leinster. It was gravel and remote back then and my sister and I had camped the other side of Sandstone and danced by the fire that night.
When we got to Leinster the next day my sister went back to Magnet and I went on down to Norseman through Kal.

Just a while before Leinster I saw some Indigenous spirits who were real upset. They were telling me that the Agnew Mine had built a road through sacred woman's ground and so I had to apologise for using the road. I notice that the fella who started the mine has had lots of bad luck regarding women in his family. This is another story too.

This was the first time I had seen Indigenous spirits talking to me in the bush.

I got struck by lightning twice within about ten minutes just out of Menzies there. It was 'green' lightning; back in those days when there was nickel in the air it would turn the lightning green. I think these days the mines are not allowed to emit that much. I started to wonder what on earth I was in for. I stopped in Kal on my way through as I had spent many years living and working there.

After a pitstop I was on my way to meet the woman who was a guardian of Njardu-Mirning b-nnah who was to become my 'boss-woman'.
She took me out bush the next day.

We set out and at the eleven thousandth unmarked tree she told me to turn left to head into unmarked country into the Nullabor desert. She 'sang' a very soft little chant and I saw an athletic and fit young full-blood ancestral messenger spirit run ahead to 'inform' the land

Guardians of our visit. It was a wiggle route and it was important that I drove exactly as she told me to so as to not damage anything 'important'. This was not a straight line journey.

After we arrived at our destination she got out of the car to 'camp down'. Before that she *sang and danced* to let the spirits know we were here. She introduced them to me. I'd hugged a few trees up to this point in my life but never before had I seen them forcefully respond back. In an instant I was convinced that Mother Earth's needs to be 'sung and danced' in the way of the Indigenous Ancestors.

Like it really needs Indigenous ceremony.
My 'boss woman' passed onto me many paintings and 'dreaming' responsibilities before she died.

On the anniversary of her death she woke me up just before the 7am news and turned the radio on for me. The station played her favorite song. I drove to work thinking she was a pretty powerful spirit who could influence radio announcers.

And one day, quite a bit later on that same song came to my ears via my CD player in my car and once again I saw her spirit sitting in my front passenger seat. As is always the case when spirits *visit* me I do not immediately question their intention for their visit as 'gwada' delivers any answers I need soon enough when the time is right.

Then the next morning I awoke to hear about a recently released movie called "Whale Dreaming" whose story apparently centers on Njardu-Mirning bannah..

One of the paintings she 'gave' and put inside me before she died was a Whale and Seal Dreaming story for that bannah. She had talked about this film but she died before it was made. I have her whale-dreaming painting. That's obviously why she woke me up. I reckon those Whale-dreamers should help with this 'dreaming' in some way.

In respect for Indigenous protocols I'll not name my 'boss-woman' but many of you will already know who she is. We had many interesting adventures out bush. And she was fascinated as to how I could possibly know so much about the bush and the medicine plants without being a 'real' blackfella.

I've also been told by quite a few people and elders that it is impossible to know many of the things I know without being Indigenous. Yes – sometimes it does make me wonder.

And it just intensifies my feeling of being a 'reverse coconut'; that is black on the inside but white (well – slightly brown) on the outside. Yes – I feel lost. Tis hard being neither a whitefella nor a blackfella. And I feel far more at home with a mob of blackfellas than with (most) whitefellas. Shame.

But I can be fairly at home either way. Like a cultural chameleon. Blackfellas so often feel like everything has been taken away and stolen from them.
And so well that it has.

The way I see it tis like the circle and the square. Circle and square story.

My boss-woman told me dreaming stories she hadn't told to any of her kin – and asked me to pass them along when the day came round. Sad really that she felt she

had to tell me who had no direct blood relationship.
I've met lots of elders who are real sad that their family
aren't interested any more.

But it's an experience that has happened to me a bit;
tribal elders telling me dreaming stuff cos I could listen
and understand and they felt that in telling me maybe I
might be able to pass certain stuff along to their kin if
and when they 'came round'. Gwada.

These poor elders feel their responsibility to pass along
their stuff and I get told it when they can't tell their mob.
Pretty dangerous position really. If I could say one thing
to young crew now it would be, **"lissen up** now".

But there is something else I really feel I've gotta tell ya
about my relationship with my boss-woman before you
make a final judgment about me. Or her even.

One of my sisters has a friend called Chris James. He's an
awesome singer and he gets around the world teaching
people to open their hearts by singing and toning.

And he used to come to Perth once a year to be the MC (Master of Ceremonies) for the New Age Expo. He would run workshops and my sister was kinda like his booking agent.

One day I was watching a video with him singing on it and my 'boss-woman' said, "I want him" so I emailed him and, much to my surprise got an answer right away. He said something like where 'n when?

Well – poor Chris, he climbed out of a small plane on that little runway at Kalgoorlie airport larger than life with his very big guitar looking real stunned like he was wondering if anyone really would meet him there. This was about three weeks after that email.

He'd met Violet beforehand and I'd told him to, "not ask her any questions nor expect any answers".

Now that's real good advice for anyone meeting an elder for the very first time. Whitefellas ask way too many questions.

Often when people ask questions they arrogantly think they know the answer already and they arrogantly are only waiting for you to let them know just that.
I think this was mentioned in an article I read somewhere or other where Chris had been interviewed. Chris told the journalist that, as a condition of meeting Violet he had been told to not ask questions.

We all started jumping up and down and acting like little screechy kids. Chris noticed us through the airport fence and figured the noise must be coming from his 'greeting party'. I think he was relieved. Anyways back to Chris later.

Lotsa blackfellas I've met don't give a shit about pieces of paper. They don't care how many degrees or certificates or graduate diplomas you might have. I explain things this way;

Whitefellas can study or research anything that takes their interest. They can get a degree and some wise and knowledgeable person full of knowledge working in a university somewhere (such as me LOL) reckons that they really do know something and gives them a piece of paper to prove it.

They can study anything and become experts. They can remember lots of stuff which might just be meaningless and useless.

Yes - this book really *is* about bush medicines but first I'm hoping you can understand a few things. See my story later on about that fella at the New Age Expo. Well – in blackfella culture you have to earn your degree each and every day of your life. You can't read a book or even ask any questions. Questions are not cool.

You have to figure stuff out for yourself and if and when you've got it you will be so informed. Gwada. If you ask a question don't expect an answer. If you're lucky you might get an answer next week – or next year. And that's only if you're finally listening and deemed ready for the answer.
You can only become the student if the teacher chooses you – and not the other way round. And now that's a deep and profound statement.

For anyone wanting a more in-depth explanation you can go here to an article I wrote at this URL here;

http://www.associatedcontent.com/article/926483/too_much_information.html?cat=37

I remember being taken out by Violet, Nuchie and Balla (R.I.P.) on 'business'. They all told me to go up along there and do 'x' and I came back to them and said, "blah blah blah". It was hilarious. They had sent me in to get rocks thrown on my head but I figured out the puzzle and passed their test.

I could tell you a lot of very funny stories. If you want to work along with blackfellas you will get tested over and over. You have to keep on earning your degree.

One day in the prison I drew a square on the whiteboard and asked them boys what it was. I said it was 'whitefella business'. Where in nature or out bush do you see anything that is a square or is even straight? Though I do now have a very precious round stone that's got *kinda* straight squares in. From Njardu bannah.

I said that without straight lines and squares there would be no whitefella culture. And that's about the size of it all really.

I think so many people out there are lost and strangled by this square culture that has no heart. And I'm unsure if there can truly be reconciliation. There seems to be fusion. And there's plenty of that. And some of it's called Indigenous politics. Or maybe we'll all have to 'go back ways'. And I reckon a lot of what people now call reconciliation is still just assimilation policy dressed in nicer looking clothes. Like look at the Intervention.

This is why ya can find lotsa poor desert people out there in remote bush who will have nothing to do with the Government – but this really is yet another story. Don't get me started on Native Title and all that stuff. Mabo and Paul Keating and Rights of Veto. Those people I'm talking about out there also avoid the so called 'benefits' of assimilation and get around real 'sorry ways'. Bit like what that girl told me about Government Departments setting up blackfellas to fail.
And it really got me round to thinking about that

'dreaming' given to me about having an earth-medicine School for Indigenous Health Workers.

The funding maybe might not come from Government and might need to come from private industry.
Or overseas even.

I'll need help to make this happen if it can – this is a big dream. And I know of maybe way over fifty Indigenous Health Workers who would straight up sign up for it. Right on the spot. And I have the academic know-how and nouse to organise to write this up.

And it is a strong testament to the inner strength and resilience of Indigenous culture that so many do want to do it despite having done those Indigenous Health Worker courses designed to help them (essentially) deliver whitefella medicine dressed up in blackfella clothes a bit.

I told some of the people at Marr Mooditj that I could do this. Violet and I did some bush medicine talks for Marr Mooditj wayback when Norm Grech was there. But the

Quartermaine woman running the show there didn't like me – I didn't have the right connections to her mind and now I notice they have been trying to use a mainstream Medical Herbalism Course Structure out there unsuccessfully.

If someone, like maybe Pundulmarra College or Batchelor pay me I would happy ways write up such a course. Probably need about $120k to get it written, framed, written up into National Training Packages lined up within the AQTF. Teacher technical lingo.

I was staying out at Burringurrah for a while. The people there didn't like the nurse and wouldn't go to her and lots sought my advice instead. Out there they called me the 'wongutha'. Maybe they recognised some njardu spirits. One of the 'boss women' from there told the matron at the Carnarvon Hospital that they wanted me to replace the nurse and the matron said she would be prepared to to this.

It would have been difficult but it made me think of the possibility that the Health System, with the right pressure from Guardians out there on isolated communities might well employ Naturopaths if they were culturally appropriate and had been trained up a bit first. They could possibly be trained up and could then provide a good adjunct to health care on isolated communities. And then they could also train up people on the communities in some basic traditional earth medicine skills.

Also around the same time one of the big 'boss men' out at Jigalong wanted me to go there and work too. When he saw all my medicines he looked and asked, "are all these made from plants". I had about two hundred or so bottles at the time.

I heard of a program too being run just out of Fitzroy there that had a pilot program happening that was using an alternative model for primary health care delivery where traditional medicine men and women were the first point of contact and doctors and nurses were somewhere further down the primary health care chain. I wonder what happened to that pilot program and I wonder what outcomes it delivered. If anyone knows get in touch with me. They actually were interested in possibly enlisting a very good friend of mine as the doctor and he attended an interview where everyone sat on the ground outside in a circle.

Desert doctor – I reckon I'll write up a few of his yarns soon as he's got some interesting ones to tell.
Maybe we might have to go back to 'old ways'. In a way or so. Culture seems to be coming back a bit some places too. I've heard people saying that.

In that video "Exile of the Kingdom" an elder stood on the Harding River Bridge above the dam and was drawing a cultural picture in the air saying, "there was my grandfather's tree, there was my grandmother's tree". Then she said, "I reckon they gotta come back to ask us how to fix things". Pretty heartbreaking.

Amazing how the Hancock regime destroyed millions of square miles of culture. And now how Gina is the richest woman and person in Australia. All that wealth originally bestowed upon a crooked Post Office worker in Wittennoom. And that too is a different story again.

I think there are Native Indian stories bit like this one too. Like when we find out we can't eat money.

I can tell you a few tales about our mining magnates and maybe I will do one day. If you wanna hear them that is. I've met some blackfellas who can tell you exactly where mineral seams are coz often they are sacred. They just might emit sacred energies used in ceremony.

Now back to Chris and that New Age Expo. Well – hold up for a minute or two.

Chris had a great time out in the desert with us all. And he told Patricia Hamilton that she needed to organise traditional openings to the Expo and acknowledge the Ancestors on the Land. He organised a Big Sing concert to help with reconciliation business. He was told he had to become a delegate for Aboriginal reconciliation as he travels the world.

My boss-woman was quite fascinated by what she saw at the Expo but she was equally upset. I had to tell someone off who had once been one of my closest friends. It was the Western Australian Flower Essences mob who were openly saying they had been given the remedies by Aboriginals.

Aboriginal politics can be such a minefield.

My boss-woman got me to do all the telling off of whoever she thought just needed it. Then my sister Carol said (that), "she loved seeing me being bossed around and me doing exactly as I was told". And I had to think about this later.

My boss-woman *never* asked me to do anything that was not in my heart or beyond my capabilities. Sure she stretched me but I never, ever got hurt by her. She only ever asked me to do what she knew I could willingly do. And I loved being bossed by her.

She was very, very good to me and one day if we can find the money required I'd like to write up her biography. She was a 'big' woman.

As we were leaving the Expo on the last day the three of us got stopped by the event organiser who asked my boss-woman, "Can you tell me about the Songlines please"?

He got impatient after about three of her sentences that didn't mention songlines and said, "I don't want to know that stuff. I want to know about the Songlines". Now this fella would definitely not have been chosen by her to be his student. He hadn't heard the term 'songlines' in her so far three sentence answer. She walked off.

Violet spoke to him for another couple of minutes and tried explaining to him that he must listen more then she walked off too. There I was left being piggie in the middle again to explain to him that he was trying to appropriate Aboriginal culture. My Aboriginal sisters tolerated and often even encouraged me talk to idiots. Maybe they thought I might be able to get through where they couldn't. But I only lingered another minute or so. He was left standing there completely puzzled.

My boss-woman always wanted me to get up in public and tell people about the time I saw the ancestral spirits responding to her.

She would tell me I was a 'real njardu' but I never felt good about saying that. I will tell people I am an adopted Njardu but some people tell me I am disrespecting her by not saying I am a real Njardu.

"You're a real Njardu Frannie", she would say. After all she had given me 'dreamings', paintings and responsibilities that she had not felt comfortable in doing with her Njardu brothers, sisters, nephews or nieces or cousins. She was busting for her nephew to "open up the b-nnah again by going through law" and she couldn't give that stuff to any old casual whitefella who happened to walk on by. I was given a skin and a skin name.

I was the 'sorcerers apprentice' to a dear friend and Indigenous medicine woman, Violet for a number of years. I once said to her, "what's the point in teaching me all this stuff"? to which she replied, "people will come to you after I'm gone". Anyways I spent another few years having very interesting adventures with her and spending any of my spare cash (earned at the prisons and at various universities) going out on adventures and on working along with her. Though she mostly works as a Nyoongar medicine woman she is also a Njardu.

And I've also, on a few separate occasions and in various places been asked by people from Wave Hill if I have mob there. And I've never been there and who knows if I ever manage to get a half decent yuledoo I might get there.

I want to go to Yulara too to get permission to publish a book I've written about Lindy Chamberlain. But that's another story. Keep posted on that one. But I reckon it will be coming out quick ways soon.

One of my younger blood sisters got claimed in Darwin by a woman whose grandfather was the same name as ours and she went out around Wave Hill and then on out to live in Borroloola for a while. People from around that way and some from Warburton are the main ones who try to claim me when they meet me but there's completely no paperwork.

There's a lot of it *about* these days.

Non Indigenous people have gotten themselves dream catchers, medicine purses and cards, totem animals, posters displaying the wise sayings of Native Indian Elders, smudge sticks, peace pipes, drums and on and on and on..... and a few people also have Australian didgeridoos or a dream-time painting or two. Australia's Indigenous mob guard their culture very jealously and do not often share important 'cultural stuff' with 'whitefellahs'.

I reckon maybe the Earth does really have an Indigenous heart and that people universally are trying to tap back into her.

But to do that I reckon it's important to understand how and why mainstream culture, in its evolution managed to so completely wreck Indigenous cultures.

Everyone should have a few Indigenous friends.

Many people these days get about '*trying*' to be green. People might buy an old art deco home in Fremantle or a patch of green in the hills and then proceed to knock down the deco home or get bulldozers to flatten the hills block and then put up a new, modern, soul-less dwelling that bears little relationship with the land it is on.

They are manage to pull apart what it is that they are seeking. Unfortunately they just 'don't get it'. This seems to hold true too for so many people who might get accused of appropriating Indigenous culture. People are seeking something but managing to dismantle what it is they are looking for.

And Yes – this book still is about bush medicine. I might be looking for a few students ;-). Or that Njardu girl I have to tell stuff to.

When we're out bush all the wild animals ignore us and get on with whatever they're doing. We don't get in their way and they don't get in ours. Though if you go out bush smelling like a hairdressing salon you just might upset the balance. Co-existence is better than taming.

A few times I've awoken to a kangaroo sniffing at the blankets, or watched goannas make their nests or seen emus 'dance' for us right in front of our eyes. And I've seen animals out there that have not been documented whitefella way. They don't come out if they're around.

When whitefellas tell stories about animals they turn the animals into people; blackfellas tell stories about the real nature of wild animals and mimic them. Therein lies an important statement too.

Lots of Indigenous folk have told me, "you know too much to be a whitefella". But I think I already said this. Seems like a lot of us might be mutta-muttas of some kind or other. Rainbow people.

I'm not greedy. I want a roof over my head that don't leak, a half decent bathroom, some new shoes and a decent car. Oh – and just now I wouldn't mind a new TV as the last one blew up a while back .And I like growing veges, watching the chooks and looking after my grannies as well as doing a bit of working. Wish someone would come and do all my housework though.

Now we're gonna get into things real soon now and I've told you my history and now you know where I come from and how I got to this spot here. Piggie in the middle.

Now let's get technical soon and really 'put a few more of those cats in with them pigeons' now.

And I'm really, really getting ready to be fired at!! From more sides than most of you can think about. And once I've written this book I won't be interested in talking to anyone who doesn't know where I come from and how I got here.

Ah – firstly I think there's another story I should also tell you's. Spiritual messages can quite often come from birds and animals and they tell you about stuff that's coming your way. Different birds and different animals depending where you are and whose land it is.

Well I usually get warnings about deaths about to happen and depending on how close the person is the gunga acts differently. The day my boss-woman died those gungas just about blocked my entrance to Canning Vale Prison. They did a big dance.

My father died without me getting any warning and I was very puzzled about this. But I soon enough got my answer.

The day after he died in Kelmscott-Armadale Hospital I pulled up at his home early the next day. As I stepped out of the car there was a feather of a large gunga just where I was about to step. But it looked real odd; the feather seemed a bit broken and it was a plastic feather! I didn't know what to make of that and thought my dad was having a lend of me and being real funny. He never seemed to like the fact that I lived and mixed and worked with Indigenous people and I think he thought I'd rejected him and his roots. I thought he was trying to pay me back in some way but was still puzzled.

I took it to sister Violet who 'read' the feather for me. She explained it well. It wasn't plastic but was the feather of a very old gunga and it was a bit broken because my father was trying to remind me not to break from my ancestral roots to him and that I must acknowledge them.

And this also explained why I didn't see a messenger gunga before he died.

Now – when in a group situation where you need to talk about your ancestral roots I wear this feather to represent my dad. And I give that explanation. And I also wear it to funerals.

Yes – he did have a laugh at me and he did teach me an important lesson.

BUSH MEDICINE

It all started last week when an elder asked me to make up some bush medicines in bulk. I haven't named her as I haven't spoken to her yet and she also ain't seen that stuff I put up on Facebook. It started when I had to develop some labels. Then I put the photos up on Facebook.

I've had support from a few corners but I also know that there are many, many people firing at me and they are all the *gone quiet* ones.

Like that girl said the other day lotsa people reckon that, "medicine is so sacred and it's the last thing left not (yet) stolen".

And if we're really, really technical about it, it has been already stolen. And it's been stolen in a number of different ways. And you need to know that I'm gonna put up a good fight to stop it being more stolen than it already is.

A bridyia who works at CALM (now DEC) told me that the Government or CSIRO 'own' (mainstream law way) every single plant on the Australian Mainland. I don't know the ins and outs of this and if I take time out to go research it then this book will be later in being available. But the information seemed reliable at the time and could just well be true.

Also – lots of people out there already – and I know a few of them – are making up bush medicines for their families and friends (mostly family). And they're making it in a few different ways. And what they're doing, if the government wanted to get shirty about things is actually illegal. Why?

Well for two main reasons; The TGA (Therapeutic Goods Act) requires that all medicines must be clearly labeled with all ingredients and excipients used in them and then a license is also needed to distribute and pass them around. So anyone already giving out bush medicines out these labels is breaking the 'mainstream' law. Now this, in itself very clearly violates Indigenous protocols. So this way the medicine has been stolen already.

They've got it covered and cornered.

And if you read my proposed labels you will see I have made a very bold statement and not named the ingredients, instead putting an acknowledgment. I had to use that acknowledgment in order to keep people off my back a bit. This puts me into a few firing lines and if I ended up using those labels and making up those medicines I too would be breaking the mainstream law if I let them be freely available and would be stirring up people from all the around the place.

The other reason is the one already mentioned before about DEC. But hang in coz I reckon there might be a good way around all this where Indigenous knowledge can be preserved and sequestered and the vestiges of some form or other of intellectual property rights retained. So, if you bear with me you will find out that I'm not actually trying to 'steal' but I am instead trying to legitimise which means save.

Now I've hung off from this now for way over 10 years.

Then there's also the issue about people like Stephen Buhner to think about too. He's an American herbalist who I believe stood on the toes of some Native Indian groups when he started using some of their sacred bush medicines. I've read one of his books and it was quite interesting and I understand completely a lot of what he was saying.

He says that the 'old ways' medicine is written into the land and that it can be channeled and put together again by intuitive sorts of people with the right knowledge. So I reckon a few intuitive sorts of folk might even get round to making up bush medicines anyways. They too might appear on the landscape sooner or later. Whether or not they get to channeling the right info is another matter. But it's not impossible.

I was out bush one day and was driving along with my 'whitefella' daughter and we stopped for a break. She roamed off into the bush and came back with a few different medicines to show me but I had never mentioned them or told her about them.

Now what I have is a license to make and dispense traditional medicines legally mainstream way. I can use different medicines from all around the world and if and when I take a suitable medical history of someone I can then prescribe anything I choose to without having to write down what is in the medicine and put it on a label. And how did I get this license?

Well in USA I would be known as a Naturopathic Physician but here in Aus I am known as an accredited Naturopath and the TGA allows me to legally dispense meds that can be ingested. Any meds as long as they are traditional 'earth-based' medicines.

IF I could get round to writing up this course (need dosh) then lotsa Indigenous Health workers would sign up. Every health worker I've met so far has said they'd like to sign up. And I reckon I could negotiate with credentialing bodies to also get people passing the course to get a TGA license after certain educational requirements were met.

Then blackfellas could make up and prescribe bush medicines. We'd probably need to negotiate some kind of clauses to do with elders and protocols. BUT the way I see the political landscape tis not quite that easy (and this is hard already).

There are other issues to address. Such as 'fusion medicine'. Now – what's that you're thinking? In a nutshell it's where two things get mixed up a bit and turn into something else entirely.

And I know quite a few people thinking they're making up traditional bush medicines that would be technically classified as 'fusion' meds. In fact most people I know who do make up bush medicine really are making up something that would be called a 'fusion' medicine even though they really reckon they are making bush medicine. In fact the only *pure and real* traditional meds I've seen so far are made deep in the Kimberley and out on (some) communities. Of course there could be others elsewhere that I don't know about.

Some people are using bushes and plants and not remembering to pick them at the right time and some get picked any old time. Then they are mixing it up in nontraditional ways and I know (from my medical training) that this changes the way they work medically in the body.

People are putting bush medicines into Vaseline – WHICH IS PETROL and, in my mind this is poisonous stuff, or putting them in lanolin – which is sheep fat or olive oil or beeswax and a whole lot of other stuff that wasn't around before the invasion. So the medicines are not working in the same way like they would have once done.

But, speaking scientifically don't let me upset the importance of the magic and the placebo as we all know about the magic in the ceremony and magic anyways. And even when put into those 'fusion ways' there's still plenty of medicine business happening.

So this is why it's called 'fusion medicine'. Mutta mutta business. And still far away from being a pharmaceutical product but that's the next rats-nest we gotta get into bed with.

Enter again the square and the circle. I think those elders, many of whom are now behind me wanted me to get this sorted.

Indigenous cultures around the world over have their own traditional medicine systems and medicine styles that I call 'earth medicines' that can be used in different cultures to bring about better health outcomes without damaging the cultural fabric very much, regardless of where they're used.

This is how I explain that pharmaceutical story and maybe I should get onto the YouTube and do a video cos, in my mind what I want to say next is real critical and very important stuff.

Friendly whitefellas come along sometimes, get taken out bush then the next thing you know they might talk to you about medicine sharing. Might seem like a good idea at the time. They're a nice fella with a good heart. But I reckon those people in the negotiating process don't understand my next point. And if they don't get to understand it then their medicines will get well and truly stolen then those blackfellas won't even be allowed to make up bush medicine traditional way themselves anymore coz they just might be breaking the agreement made with the pharmaceutical company.

And after that they won't even be allowed to make it up fusion ways either. It'll all be by pharmaceutical license only.

Then the only person I know of who might be able to help you if and when you get upset is Robert Eggington at Dumbartung. But this is another story I'll explain to you after explaining how these pharmaceutical people work. I just noticed Dumbartung has a Facebook page. Robert understands how these pharmaceutical mob work.

OK now – imagine you're walking out bush with one of your whitefella mates and you mention, "ah – now there goes a good bush to cure cancer". This whitefella gets real interested and you let him take some home to make up a pot of tea but the whitefella (who might be a fella with a good heart who also just might have good intentions to cure the world of cancer) takes the bush around to his chemist friend.

Now there is a widespread rumour about that says there are lots and lots of medicines to yet be found and that the Indigenous peoples of the world know about then and use them and there's lots of hungry chemists who want to uncover this information who also might have good intentions about curing millions of people. I haven't got much beef with them but it's the way that the whole pharmaceutical industry works that I'm very wary of. Now let me say again real clear ways that they are not going to get any information from me that would help them in identifying these plants. And I don't know the Latin names for the bush medicines I know about anyway but I have noticed a few tourism type blackfellas using those Latin names and that I would say might be a culturally very dangerous thing to do.

Where those chemist fellas are gonna manage to get the info from just might be you; accidentally. And cos you don't understand the nature of the pharmaceutical beast. It's like the story about the frog and scorpion or the circle and square.

Back now to your mate who got enough bush to make a pot of tea who took it around to his chemist mate. That chemist got intrigued and most likely excited. He knew his mate reckoned it might cure cancer so with this in mind he looked down some real powerful microscope. There he saw a zillion parts cos plants do have lots of parts.

But being a chemist tarred by the pharmaceutical brushes he had to decide which chemical was the most important one. Yeh – it is true that lots of medicines these days came first from plants but let me tell you really clearly that this is not entirely as true as it might seem. Some chemist in the waybacks did look down a microscope and chose one chemical.

But I think I've got to go sideways again to better explain myself so we'll come back to those chemists again in a little while. Plants are pretty smart things and they are smarter than those chemists who see parts of plants as isolated chemicals.

Bit like a dance group I guess. If you take one dancer out and expect him to be able to do the job of the whole dance group he wouldn't be able to pull it off. The part he plays in the scheme of things for the dances doesn't make too much sense without the context added by the whole group put together.

So whilst a whole plant, straight from the raw might be able to do a pretty impressive job, so might only one chemical which is then turned into something patentable such as a chemical analog not do the same thing as the whole plant.

Plants with toxic parts can sometimes and often be not toxic when the whole plant is used because; plants have buffering systems and this allows the body to assimilate what it needs from the plants.

This is a pretty simplistic way of describing it as this topic alone could take up a whole, big and complex book. But I reckon you's should get the idea.

If the same standards were used for assessing our foods as are applied to these pharmaceutical drugs then we'd be all saying, "What – no potato"?

And quite a few common foods would be off the shelves too. Things can go a bit skewy when those chemists gets into the ring. And don't get me wrong either – we all know that plants are real poisonous.

OK - a lot of earth based traditional medicine systems from around the world that I know of do have certain things in common and nearly all of these systems usually have one central plant or bush that is the most sacred and used the most often.

Now I need to explain some medical herbalism terminologies here. So I can make another important point. There's ALTERATIVES and AMPHOTERICS and TONICS.

These are 3 very important words that are also concept words. And just about all of those central plants (in various cultures) that I am talking about have all of these three properties.

An *alterative* is a broad spectrum healer that is a blood cleanser that can often clear up a whole range of conditions but an alterative by itself is not as good as a plant that has all three of those properties mentioned above.

A *tonic* – well you can probably work that one out for yourself. It boosts the bodies energies, boosts the immune system and provides energy.

An *amphoteric* – well this is an interesting category of plants and is quite complex and absolutely no pharmaceutical product or formulation ever is one of these. Amphotericity belong only to world of healing plants.

I'll tell you about garlic which is an amphoteric. So that you can get the idea about what I'm talking about.

When garlic is prepared in a medicinal way you can take it when you have blood pressure problems. If your pressure is high it will lower it and if your pressure is low it will raise it. So an amphoteric is like a smart plant that adjusts a bodily organ system. And there are many amphoterics that work on different systems within the body.

In traditional ways of plant medicine and herbalism alteratives are the first stop in a treatment regime and are usually taken for at least a month. During this month many ailments will clear up but sometimes there are others left that need to be treated with much more specific medicines after you have taken the alterative for a couple of months.

Many Indigenous people, when I talk about bush medicine think I'm talking only about *the* one central plant. But I know about many, many other plants out there in the Aussie bush that have medicine properties that can be used either along with the one that is the alterative/amphoteric/tonic or separately for a very wide range of conditions.

My boss-woman wanted me to start up a school, make up the medicines in like a medicine factory, get organised to get some of her people licensed, get medicines out to cure so many of the ailments and to (also)pass along my pretty vast knowledge of medicinal plants.

I know enough to have a pretty broad Indigenous Pharmacopeia that could treat a pretty wide range of ailments. We'd end up with just like an Indigenous chemist shop.

And sure – there are many parts of Australia whose plants I am unfamiliar with too but with a network of fellas trained up to get their TGA licenses traditional medicine could come back in a big way and without ever getting stolen all of the way.

Then communities could also sell their medicines to other practitioners or beyond pending certain conditions. Now who dares to now to tell me I am trying to steal sacred knowledge? And does there happen to be anyone else out there in Aus who is Indigenous (or at least adopted as one), has studied naturopathy, has a TGA license, a lot of traditional med' knowledge and who also knows how to write up course content in a culturally appropriate way? I might just be unique after all. And have a special job to do.

Now we still need to get back to the chemist fella who has to decide on only one chemical in the plant? And why does he have to choose only one you might be wondering. Well – it's the name of the game.
Welcome to the pharmaceutical industry. He chooses one part of the plant (often an alkaloid) and – zippo – makes what is called a chemical analog. And a chemical analog can be *patented*. It's basically like plastic. Capitalism dressed up in quality controlled clothes. Not to mention globalism.

So now the pharmaceutical company has gotten themselves a chemical analog that is entirely owned by them. The drug they make from it has a chemical in it that *resembles* something that was once part of a plant.

But – and this is a very big but – it doesn't work in exactly the same way as the plant itself does. And anyone half smart would be able to work that one out. And also remember now those dancers.

It cannot be an alterative or a tonic and it most definitely cannot be an amphoteric. It can now cause many side effects which the plant medicine, made up using traditional methods (not fusion methods) would not do.

The pharmaceutical product is also now 'quality controlled' (another broad topic) and can now be dispersed and prescribed all over the world without anyone ever having to prune your trees again. So what now replaces the traditional medicine product is now a product that will not work exactly like the traditional one. It can be much more dangerous and to my mind it also becomes far less effective.

If a community you were part of was manufacturing traditional medicines; picking them at the right time of year and preparing them according to traditional methods your community might be able to develop a very viable business selling traditional medicines and also making them up for your mob to improve their health.

But if only one person on one community accidentally or knowingly sells a patent to a pharmaceutical company then it's all over red rover. That one person or community might get some form of benefit but there would be no benefits that could be spread out and shared around.

And lots of medicines that are in one place are in many, many more other places too so if one mob miles away manage to negotiate with a pharmaceutical company then only one community might reap any gains from royalties from that plant available in many places. This story vaguely resembles the Native Title story where not everyone gets benefits and this can causes lots of acrimony and divisions and jealousies.

By making medicines up traditional ways and not selling any intellectual property rights to a chemist and a pharmaceutical company then much wider mob can benefit. And have jobs too. And much, much better health outcomes.

A lot of those old plants, I reckon have properties that could cure common afflictions such as diabetes and kidney disease. And eye problems and all kinds of things.

I just wanna get this ball rolling along and also maybe get it moving beyond the central alterative plants.
Now there is a story a lot like that mate we talked about before who took his chemist friend out bush.
Here in WA's south. And CSIRO got hold of it, looked down their microscopes and went all the way along to developing a patented chemical analog. They got pretty excited about it and scientific papers started coming out.

There was a lot of buzz.

But the fella who gave his friendly whitefella mate enough bush to make a pot of tea got a bit upset when all this scientific buzz started happening. I guess he felt cheated and like something important and sacred had been stolen from him with no regard to his knowledge and he got real upset. Like his knowledge had been *misused*.

Perhaps if people were just gonna make up for themselves a few cups of bush tea he may not have minded quite so much. He didn't realise the nature of the beast that is the pharmaceutical world.

And it reminds me about so called 'nature identical' foodstuffs. Anyone who has tasted 'nature identical' (NI) vanilla can tell you that it doesn't taste like the real thing. So those chemists peering down their microscopes do obviously miss a few things or two.

I don't know all the ins and outs of this next story but somehow seem to recall that Robert Eggington of the Dumbartung Cultural Corporation took on the CSIRO for misuse of Intellectual Property Rights and appropriation of Aboriginal Culture when it was heard that a pharmaceutical patent might arise as a result of that fella giving his mate enough bush to make a cuppa. And the case may have even gone (or at least it threatened to go) all the way to the High Court of Geneva. It was resolved, I believe in favour of Aboriginal Cultural rights and CSIRO ceased any research or interest in the plant.

So what I see as a viable solution is to get TGA certificates for Indigenous Medicine people so they can then dispense traditional medicines.
Well I reckon I'm just about done now with my story and I reckon that I've said everything I need to say.

Thank God for that. Let's see what happens next. Let's see who comes out of the closet cos I'm sure having a good go here in trying to wake them up to come out to talk to me.

If you feel you really need to talk to me then arrange to do so.

But maybe I just now need to precis what this little book is basically about or what I would like to see come about; which is that Australian Indigenous traditional and bush medicine knowledge (as it is) can be preserved and shared around in a way that doesn't steal from or appropriate culture and that also benefits the primary physical and spiritual health of the Aboriginal Nation of Australia.

I've drawn up now what I see as being the 'mud-map' so lets get round to making it happen now.

*DEDICATED to all those 'behind' me.
And to the ones alongside.
Also to those who, one day I will be behind.*

PIGGY IN THE MIDDLE STORY-

About Bush Medicine

Jaahda Jinnah